I0814155

Life Unplugged

TOP OF THE WORLD

OFF-GRID MOUNTAIN LIVING

by Mari Bolte

full tilt PRESS

Full Tilt Press
42964 Osgood Road
Fremont, CA 94539
readfulltilt.com
Full Tilt Press publications may be purchased for educational, business, or sales promotional use.

ISBN: 9781629209562 (hardcover)

ISBN: 9781629209722 (ePUB eBook)

ISBN: 9781629209401 (PDF eBook)

Editorial Credits
Editor: Jill Kalz

Copyeditor: Kristin Russo

Designer: Sara Radka

Image Credits
Cover: ©Mumemories / Getty Images; page 1: ©Pgiam / Getty Images; page 3: ©MaryDesy / Shutterstock; page 4: ©C-mere / Getty Images; page 4: ©Wgcwriter / Wikimedia; page 4: ©Emil Timplaru / Shutterstock; page 7: ©massimo colombo / Getty Images; page 7: ©Raphye Alexius / Getty Images; page 7: ©fhm / Getty Images; page 8: ©Marco Bottigelli / Getty Images; page 9: ©Bartosz Hadyniak / Getty Images; page 11: ©massimo colombo / Getty Images; page 13: ©Geir Pettersen / Getty Images; page 14: ©Monica Bertolazzi / Getty Images; page 15: ©Feng Wei Photography / Getty Images; page 16: ©KenCanning / Getty Images; page 17: ©Troy Harrison / Getty Images; page 19: ©Devesh Gautam / 500px / Getty Images; page 21: ©Jakob Radlgruber / EyeEm / Getty Images; page 23: ©kafka4prez / flickr.com; page 24: ©Mihai_Andritoiu / Shutterstock; page 25: ©EBCI / energy.gov; page 27: ©Christoph Wagner / Getty Images; page 29: ©Carmian / Getty Images

Printed in the United States of America.

CONTENTS

INTRODUCTION

Off the Grid

Are you ready for an adventure? Pack a bag and unplug!

Today more than 1.7 billion people around the world live off the grid. To some people, this means not relying on **public utilities** at all. For others, it means living an Earth-friendly lifestyle. It is a way to live more independently. The only person you rely on is you.

public utility: an organization that manages public goods and services, such as a water or power company

TOP MOUNTAIN RANGES AROUND THE WORLD

Some off-gridders don't have a choice. They may live in a place without public utilities such as water and electricity. But others go off the grid on purpose. They might want to reduce their impact on the planet or be more independent. They might want time away from technology.

People who choose this lifestyle find or make their own power sources. **Solar**, wind, or **hydropower** are used instead. People collect and store rainwater. They grow the food they eat. Waste is disposed of without sewer services. Outhouses, **septic tanks**, and **composting** get that messy job done.

Not every person who lives off the grid does it the same way. But each one learns lessons as they go. Find out what it takes to live life unplugged—in the mountains!

---

solar: having to do with the sun

hydropower: electricity created by fast-moving or falling water

septic tank: an underground tank that holds sewage and wastewater

composting: recycling food and plant waste into nutrient-rich organic matter

CHAPTER 1

Welcome to the Mountains

Nearly one quarter of Earth's land is covered in mountains. Mountains usually stand at least 1,000 feet (305 meters) tall. Most of them have a sharp **peak**. They can stand alone or be part of a larger mountain range.

Millions of people live on mountains. Some live in cities that grew large in a mountain's shadow. Park City, Utah, and Mexico City, Mexico, are examples. Other people live simply, in small mountain villages.

People have been living unplugged since the beginning of time. Which off-grid mountain home on the next page would you choose?

FAST FACT

Height and depth on Earth are measured by sea level. The Dead Sea is the lowest place on Earth at 1,414 feet (431 m) below sea level. The top of Mount Everest is the highest place on Earth at 29,029 feet (8,848 m) above sea level.

---

peak: the pointed top of a mountain

Log Cabin

PROS:

- lots of building materials available
- permanent
- eco-friendly
- traditional

CONS:

- not fireproof
- difficult to use machinery to help build
- pests can get in

Tree House

PROS:

- great views
- tree protects the home
- limited size can help control expenses if the plan is simple

CONS:

- adding things such as running water can add cost
- need an old, strong tree that will live a long time
- limited size

Cave Home

PROS:

- natural shelter
- naturally insulated
- eco-friendly

CONS:

- dark inside
- moisture can be a problem
- lacks ventilation

The amazing views are one reason off-gridders choose to live in the mountains.

Those who live in the mountains like the views. They enjoy the fresh air. The mountains help them feel independent. But living at higher **altitudes** comes with some challenges. The higher you go, the thinner the air becomes. Anything above 8,000 feet (2,400 m) above sea level is considered high altitude. Here, the **air pressure** drops. It is much colder. People who spend a lot of time at high altitudes can find it hard to breathe deeply.

altitude: the height of an object above a surface

air pressure: the weight of Earth's atmosphere

Finding land to buy in the mountains is easy. Building a home there is harder. Things such as electricity, water, or roads can be few and far between. Hauling building supplies up and down mountains is difficult. Making sure homes are level can be a challenge when the ground is frozen for much of the year. Freezing weather limits when building can take place too. Clearing away snow takes time and costs money. And not everyone likes working outside in the cold.

LIVING HIGH IN THE SKY

People who traditionally live at high altitudes have **adapted** to their environment. Their bodies have changed over time. In some cases, their blood carries more oxygen. In others, they breathe more often to make up for the lower levels of oxygen in the air. Some people living in the Andes Mountains and Tibet have these adaptations.

---

adapt: to become adjusted

High-altitude locations have cooler seasonal temperatures. In the Rocky Mountains, the average summer temperature is about 75 degrees Fahrenheit (24 degrees Celsius). The food-growing season is short. Mountain dwellers must pay attention to the direction in which the slope faces. If the slope faces east, it will get morning sun only. If it faces south, it will get light all day.

Gardening in low meadows and ridges protected from wind can be a good choice. The soil there is richer, and the ground is flatter. But the air is also cooler. Gardening in raised beds is a way around frozen soil. There are also many seeds designed to grow well in high elevations or cold temperatures. The best choices for mountain gardens are cool-weather leafy greens, root vegetables, and flowering vegetables, such as broccoli.

Mountains are a great natural source of food if you know where to look. Nuts and seeds have lots of **nutrients**. Acorns, walnuts, beechnuts, and pine nuts are a few examples. Apples, crabapples, and berries can be found in valleys and on hillsides. Using good sense when **foraging** is key. If you don't know what something is, don't eat it. Use a field guide or check with an expert.

FAST FACT

Some off-gridders collect rainwater in barrels. Others dig their own wells. Sometimes, the only way to get water is to haul it in from long distances.

---

nutrient: a substance that provides nourishment for living things

forage: to search for food in the wild

The temperature drops roughly 1 degree for about every 300 feet (90 m) you move up a mountain.

CHAPTER 2

Danger in the Mountains

Wind, snow, and other dangers can make mountain living a real challenge. Mount Washington in New Hampshire is the highest peak in the northeastern United States. It has some of the most severe weather in the world. Hurricane-force winds blow for one third of the year. Snow, ice, and sub-zero temperatures do not make it an inviting home. But some people do live there! Weather observers and park rangers are there year-round.

FAST FACT

Altitude sickness is also called hypoxia or mountain sickness. It can be deadly at altitudes above 25,000 feet (7,600 m). Headaches, trouble sleeping, vomiting, making poor choices, and being very tired are signs of the illness.

Avalanches are dangerous. Snow falls onto mountaintops, creating layers over time. The bottom layer is called snowpack. It grips the mountain and keeps all the snow from sliding off. But if the snowpack is weakened, the layers come tumbling down, burying everything in their path.

Even with today's technology, no one knows exactly when an avalanche will happen.

Avalanches include blocks of ice, clouds of thick snow, trees, and rocks. They can reach speeds of 80 miles (129 kilometers) per hour in five seconds. Maximum speeds of 200 miles (322 km) per hour are not uncommon. If you're not crushed by the snow, you can die from lack of oxygen beneath it. Every year, roughly 150 people are killed in avalanches around the world.

CLIMATE CHANGE

Climate change threatens everyone, but people in the mountains will feel it first. In the northern part of the planet near the Arctic, glaciers are melting. These large ice sheets are several hundred to several thousand years old. About 10 percent of the land on Earth is covered by glaciers. But a warmer planet means that ice is melting faster. The water can flood crops, towns, and houses. Animals that are hunted for food move farther north, looking for colder climates. Entire **ecosystems** are affected.

Humans trigger 90 percent of all avalanche disasters.

Climate change makes mountain living more and more dangerous. In the winter, snow falls and is stored on mountaintops. In the spring, it melts, releasing water into rivers and lakes. As the planet's average temperature rises, less snow falls, and the snowpack melts earlier every year. The summers are longer. The dry conditions can feed wildfires. Rain instead of snow can cause rockslides and avalanches. It can also cause flash floods.

---

climate change: a change in global or regional climate patterns

ecosystem: all the plants and animals in a linked environment

At the lowest level, mountains are covered with trees. These trees are usually deciduous, which means they lose their leaves in the winter. The lower forests often blend into other types of flatland ecosystems, such as deserts, rain forests, or **tundra**.

As you move up the mountains, trees thin out. They change to tall coniferous trees, such as pines and evergreens. The path becomes sharper, and moss-covered cliffs make getting around more difficult. It can be easy to fall or seriously injure yourself. Help may be many miles away.

More than 85 percent of the amphibians, birds, and mammals on Earth live in the mountains.

tundra: a wide, treeless area where the ground is permanently frozen

The views are more beautiful the higher you go, but it is harder to find food. Some people solve this problem by raising their own animals. Chickens, cows, sheep, and goats all do well in the mountains. But they need to be kept safe from predators, such as coyotes, wolves, mountain lions, and bears. Deer and elk may eat the hay, corn, or other food stored for the farm animals.

Sharing living space in the mountains with bears and other wild animals can be tricky.

FAST FACT

Giving wild animals their space is the best way to avoid being attacked by them. Be mindful of your surroundings to avoid surprising them. And don't feed or try to pet them, no matter how cute they look!

CHAPTER 3

Rewards of the Mountains

Some of the most beautiful places on Earth are part of the mountain ranges that stretch across and over the land.

Denali is the highest peak in North America. It is part of Denali National Park in Alaska. Five Native groups have lived on the land around the mountains for generations. They catch salmon and other fish in the summer, and hunt moose in the fall and winter. Thirty-nine different species of mammals call the mountain home. Beautiful views of the Aurora Borealis can be seen from August to April.

Harsh weather conditions don't bother the animals who live there. Some have adapted to stay warm during long, cold winters. Others can scale steep cliffs. Learning how these animals live can be great inspiration for going off the grid yourself.

FAST FACT

The Aurora Borealis is also known as the Northern Lights. Electrically charged particles from the sun enter the layer of gases around Earth and create rippling, colorful lights.

The most common color for the Northern Lights is green.

The Alps are the highest and largest mountain range in Europe. They touch 11 different countries. And they are the most inhabited mountains in the world. But that doesn't mean they're crawling with people. Going off-grid in the Alps can be an adventure.

People who live in the Alps can walk, rock climb, ski, kayak, snowshoe, and swim whenever they want. There are many spas, for people who would rather relax than hike. Some places, like Lake Cauma in Switzerland, are said to have healing properties. The fresh air is good for the lungs. The water is clean and delicious. Living away from city sounds can be soothing.

WEEKEND WARRIOR

Give off-gridding a try for a weekend. The mountains may be far away, but you can imagine your own. Try camping outdoors. If you can, set up your campsite on ground that's not flat. You will have to carry your supplies to your sleeping area. And make sure they don't roll back down the "mountain"!

Humans have called the Alps home for 50,000 to 60,000 years.

CHAPTER 4

Who Lives in the Mountains?

The Earthaven Ecovillage is near Asheville, North Carolina. It is between 2,000 and 2,600 feet (610 and 792 m) in the mountains. The community was founded in 1994. It covers 320 acres (129 hectares). Today there are around 100 people who call the village home. The land they use belongs to the Catawba and Tsalagi people, also known as Cherokee. The village works in partnership with them to support ecology and live **sustainably**.

EARTH-FRIENDLY LIVING

There are around 10,000 ecovillages around the world. Some are traditional rural villages. Some, like the Earthaven Ecovillage, are intentional, which means they were created by people with a shared vision. Urban ecovillages exist within cities such as Los Angeles and San Diego, California. They focus on sustainable living.

Most Earthaven Ecovillage homes use south-facing windows to collect as much sunlight and heat as possible.

Homes are built from natural or recycled materials. They are on slopes, which leaves all the flat space for growing food. A natural spring and a huge storage tank provide water for the community. Solar panels provide power.

---

sustainably: in a way that's able to meet our needs without completely being used up

Just a short journey west lies the Eastern Band of Cherokee Indians (EBCI). They call the southern Appalachian Mountains home. Nearly 13,000 **enrolled** members are part of the EBCI. In 1996, they re-took control of the land where their ancestors once lived. It was the first time Cherokee people controlled the land in 175 years. And they have moved quickly to live in Earth-friendly ways.

The town of Cherokee, North Carolina, is the governmental center for the EBCI.

FAST FACT

Indigenous people have the right to form their own governments on their own lands. This is called tribal sovereignty. In the United States, there are 574 tribal nations.

The EBCI's solar farm contains more than 2,000 solar panels.

Native people are the original off-gridders and caretakers of the land. In 2008, a new school was built with eco-friendliness in mind. Local trees were used for the interior woodwork. Around 90 percent of the construction waste was recycled. The **geothermal** heating and cooling system relies on nearly 300 wells that pull air from the earth as needed. The school is water- and energy-efficient.

For years, the EBCI has vowed to work toward a natural, sustainable environment. They want to use alternative energy and reduce the total amount of overall energy used. In 2020, they built a solar farm to cut down on energy costs in the community. By collecting their own energy, they also do not have to rely on anyone else. The EBCI is truly taking off-gridding into the 21st century.

---

enrolled: recognized; tribal enrollment is when an individual is a member of a federally recognized tribe

geothermal: heat produced by the Earth's interior

Living life unplugged in the mountains can help you feel closer to nature. The views of the sky, being among the trees, and relying only on yourself can be freeing. Working outside in the fresh air is good for the mind and the body. And caring for Earth is never a bad thing.

Not everyone can make living off the grid work. It's a full-time job, not a camping vacation. But even if you can't go off-grid, there are ways to make your carbon footprint smaller. Be mindful of how much water you use. Use things made of recycled material, and then recycle them in turn. Eat locally grown food or grow it yourself. Caring for the planet is something everyone can do.

TAKE ACTION!

Climate change is a real problem. It will only continue to get worse unless something is done. Learn about the challenges climate change is causing. Think about the future. Consider a career in Science, Technology, Engineering, or Math (STEM). You can be a scientist or a geologist or an inventor. Maybe you will have a great idea that will make a difference!

Despite the challenges of living there, people will always find ways to go off-grid in the mountains.

QUIZ

1. Cave homes may have moisture problems.

   A. True

   B. False

2. What is *not* a sign of altitude sickness?

   A. difficulty breathing

   B. super strength

   C. tiredness

   D. headaches

3. What makes avalanches so dangerous?

   A. fast speeds

   B. debris, such as rocks and trees

   C. weight of the snow

   D. all the above

4. What is the highest peak on Earth?

   A. Denali

   B. Mount Washington

   C. Mount Everest

   D. Matterhorn

Answer Key: 1) A; 2) B; 3) D; 4) C

ACTIVITY

ANIMAL ADAPTATIONS

Choose a creature that lives in the mountains. Research where it lives. What does it eat? How does it stay warm in winter? Choose one of its adaptations. Then think of ways that adaptation could help humans living off-grid in that same environment. Challenge yourself to design a new product that uses that adaptation.

ANIMAL IDEAS

- wood frog
- mountain goat
- eastern spiny softshell turtle
- golden eagle
- lynx
- wood bison
- bull trout
- cougar
- gray wolf
- grizzly bear

GLOSSARY

adapt: to become adjusted

air pressure: the weight of Earth's atmosphere

altitude: the height of an object above a surface

climate change: a change in global or regional climate patterns

composting: recycling food and plant waste into nutrient-rich organic matter

ecosystem: all the plants and animals in a linked environment

enrolled: recognized; tribal enrollment is when an individual is a member of a federally recognized tribe

forage: to search for food in the wild

geothermal: heat produced by the Earth's interior

hydropower: electricity created by fast-moving or falling water

nutrient: a substance that provides nourishment for living things

peak: the pointed top of a mountain

public utility: an organization that manages public goods and services, such as a water or power company

septic tank: an underground tank that holds sewage and wastewater

solar: having to do with the sun

sustainably: in a way that's able to meet our needs without completely being used up

tundra: a wide, treeless area where the ground is permanently frozen

READ MORE

Doeden, Matt. *Can You Survive Hair-Raising Mountain Encounters?: A Wilderness Adventure.* Mankato, MN: Capstone Press, 2022.

Huddleston, Emma. *Outdoor Expedition Bucket List.* Minneapolis: Abdo Publishing, a division of Abdo, 2022.

Payne, Stefanie. *The National Parks: Discover All 62 Parks of the United States.* New York: DK Publishing, 2020.

INTERNET SITES

Appalachian Trail Conservancy: Native Lands
https://appalachiantrail.org/official-blog/native-lands/

National Geographic Kids: Mountain Habitat
https://kids.nationalgeographic.com/nature/habitats/article/mountain

National Park Service: Animals, Plants, and Habitats
https://www.nps.gov/subjects/mountains/animals-plants-and-habitats.htm

INDEX